Design Thinking Notebook

Design Thinking Notebook

A step-by-step process to help innovators create solutions to everyday problems

iSTEM Publishing

By accessing this book, you accept this disclaimer in full.

The information contained within this book is strictly for educational purposes. Apply ideas contained in this book at your own risk. Your results are likely to differ than those of the author.

The information in this book is based on the author's knowledge, experience and opinions. The methods described in this book are not intended to be a definitive set of instructions. You may discover other methods and materials to accomplish the same result. Your results may differ.

Although the publisher and the author have made every effort to ensure that the information in this book was correct at press time and while this publication is designed to provide accurate information in regard to the subject matter covered, the publisher and the author assume no responsibility for errors, inaccuracies, omissions, or any other inconsistencies herein and hereby disclaim any liability to any party for any loss, damage, or disruption caused by errors or omissions, whether such errors or omissions result from negligence, accident, or any other cause. The information is provided "as is," to be used at your own risk.

© 2024 iSTEM Publishing. All rights reserved. Including the right to reproduce this book or portions thereof, in any form. No part of this book may be reproduced or transmitted in any form or by any means, electronic or mechanical, without written permission from the author, except as permitted by Canadian copyright law.

No part of this publication may be stored in a retrieval system, or transmitted in any form or by any means, electronic, mechanical, photocopying, recording, scanning, or otherwise, without the prior written permission of the author.

For permission requests, contact: istempublishing@gmail.com

All referenced trademarks are the property of their respective owners.

Published in Canada. Cover and template design © 2024 iSTEM Publishing.

PAPERBACK ISBN: 978-1-7382348-0-6

*For Phoenix and India,
the inspiration for all that we create.*

Design Thinking Notebook

A step-by-step process to help innovators create solutions to everyday problems

About This Book 1

Design Number 1 3

Design Number 2 21

Design Number 3 39

Appendix: Extra Build, Test & Evaluate Pages 57

About This Book

Using design thinking to create solutions to everyday problems, is a method being introduced to students of STEM related subjects throughout the world. As a teacher, it has been challenging to find resources that enable students to follow the design process, with ease and success. Through many iterations, the following templates have been developed and proven to be successful, in both a classroom setting and also with my own children at home. Having a template with guiding information, has enabled the people using these, to follow the design thinking process with ease and create amazing products as a result.

Here are some projects that have been developed using the templates from the Design Thinking Notebook:

Trash Bash Project - Try to use trash found around the home to create something that is useful.

Robot Designs - Design a robot to complete a challenge or series of challenges. Robotic competitions such as VEX Robotics, FIRST Lego League (FLL), FIRST Tech Challenge (FTC) and the FIRST Robotics Competition (FRC) are just a few platforms to explore. For VEX and FLL, once you purchase the hardware to build the robot, they provide free online lessons and resources to get you started.

Pinewood Derby Car - Design a pinewood car to be raced against others. Cars are released from the top of a track and travel approximately 13 meters to the finish line. Cars only use the potential energy from the height of the track and aerodynamics from the car design to reach top speeds. Pinewood cars come as a standard wood block that can be carved into various shapes easily.

CAD Design and 3D Printing - There are several CAD projects that can be created using free online software. Autodesk TinkerCAD, Solidworks and 3D Slash are a few to try for people first starting out. For more advanced users Autodesk Fusion 360 and SketchUp can be used. All are available to use for free and include online tutorials to get you started. These softwares enable you to create 3D objects that can be 3D printed. Projects include; key chain designs, jewelry designs, car designs, and more. Additionally, the organization Makers Making Change focuses on creating assistive devices using 3D printers and always has design challenges available from actual users.

Game Design - When thinking through developing a game produced in Scratch, Python, Java, Unity or any other language, working through the design thinking process is a great tool to develop ideas and create something original.

As you can see, there are so many possible ways to use this Design Thinking Notebook. As you go through the process of design thinking, you will gain a greater understanding of how the process works and see its benefits through what you create. By encouraging creativity, we hope to help you learn and grow as a maker, inventor and innovator.

DESIGN PROCESS

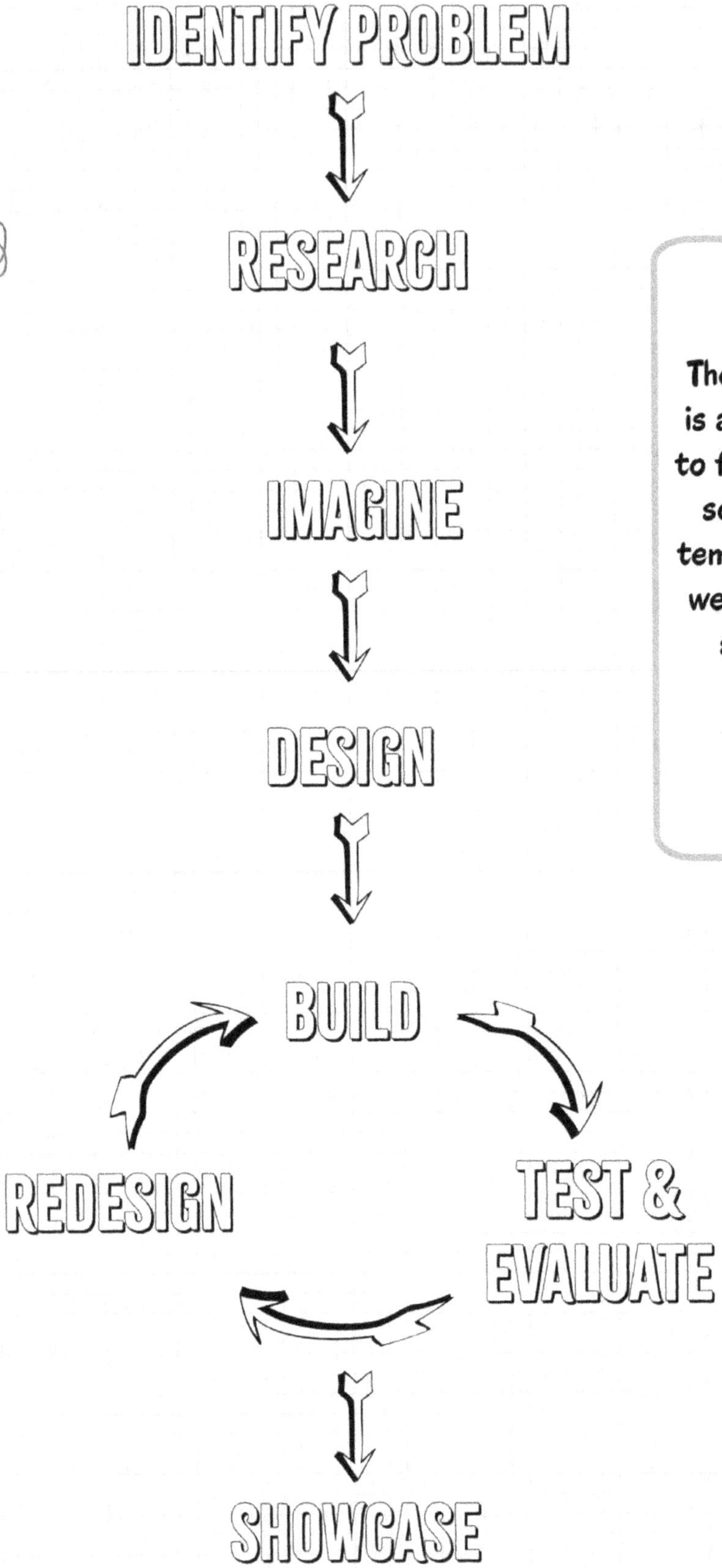

IDENTIFY PROBLEM
↓
RESEARCH
↓
IMAGINE
↓
DESIGN
↓
BUILD → TEST & EVALUATE → SHOWCASE
↑_____↓
REDESIGN

The design thinking process is an easy to use framework to facilitate creative problem solving. By using planning templates to generate ideas, we hope to help you to find solutions to real-world design challenges.

DESIGN NO.1

DESIGN PROCESS

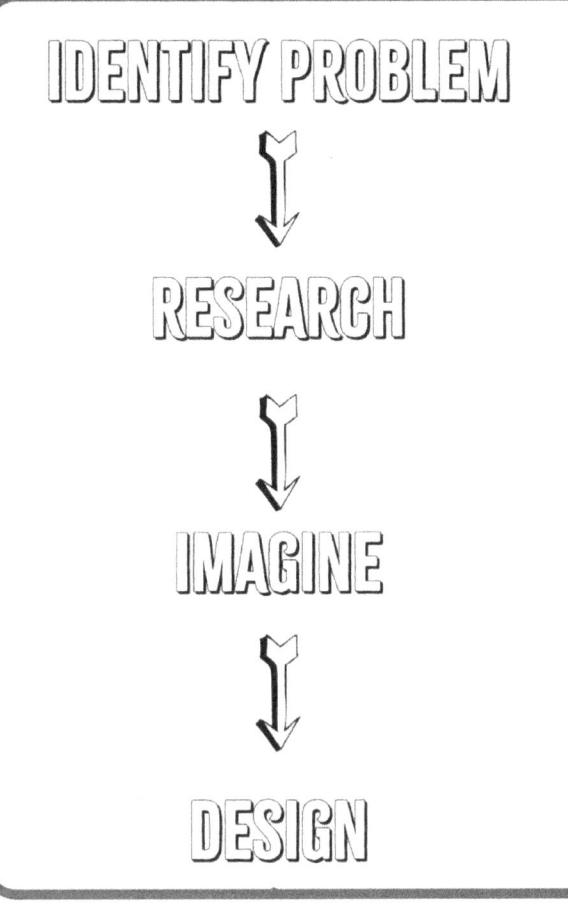

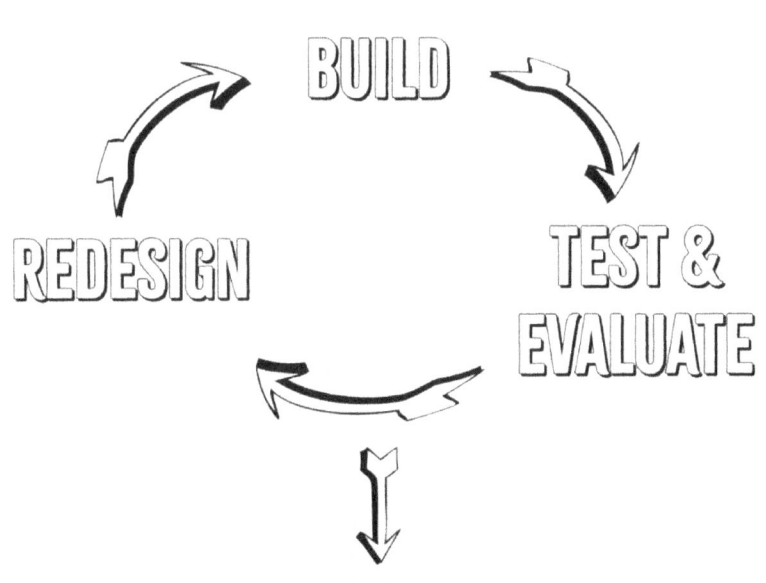

Exploration part of Design Thinking

In this section, dive deep into a possible **PROBLEM** you would like to solve. You might look at an object and ask yourself what more can be added to make it better? You may have a design **PROBLEM** that exists and you want to find a design solution for it. Or you may just want to invent something really cool! Once you determine a **PROBLEM** (**OPPORTUNITY** to create) you need to think about this further.

1 IDENTIFY THE PROBLEM

IDENTIFY THE PROBLEM

Identify the problem you want to solve.

NEEDS

After identifying your **PROBLEM (OPPORTUNITY)**, start to think about what the design (solution) you are creating **NEEDS** to have. These needs might be thought of as requirements. What does the design have to have in it? Such as: a certain size, characteristics, or something it must do.

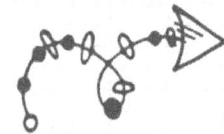

LIMITATIONS

Consider what **LIMITATIONS** you have in creating your design. Do you have a limited amount of time? Or only certain supplies or materials? This list will help you design within your limits while ensuring you include everything that is required.

RESEARCH

WHAT DID YOU FIND OUT?

RESEARCH to see what others have developed and created related to your **PROBLEM**. Here you might discover things you like about other people's designs and get inspiration for what you might include in your own design. You might also find things from other designs you do not want to include. Make notes on your research here.

IMAGINE, IMAGINE..

During **IMAGINE** you will come up with ideas for your design. The goal here is to sketch out as many design ideas as possible, not worrying about whether the idea is good or bad.

When you have imagined and sketched out as many ideas as you can, you will list the things you **LIKE** and **DISLIKE** about each your design ideas. Try comparing your ideas to come up with these.

IDEA 1:

IDEA 2:

LIKES:

DISLIKES:

LIKES:

DISLIKES:

...... IMAGINE!

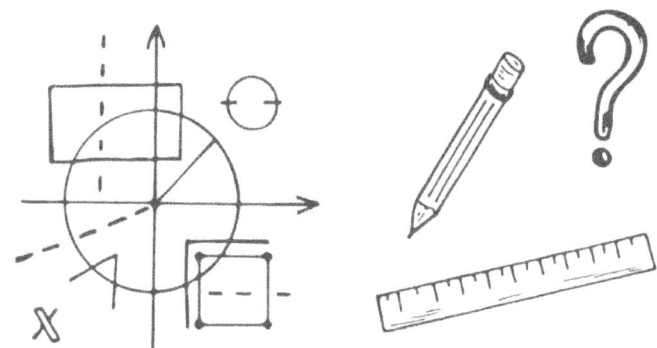

IDEA 3:

IDEA 4:

LIKES:

DISLIKES:

LIKES:

DISLIKES:

FINAL DESIGN

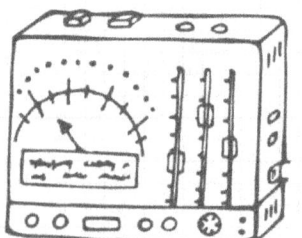

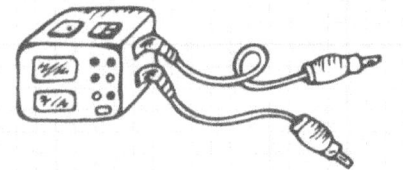

Here you will come up with a **FINAL DESIGN**. Try to incorporate as many of the ideas you liked from the **IMAGINE** section. In your **FINAL DESIGN**, try to add in as much detail as possible, such as labels and explanations, so you know exactly what you want to create and build.

CHECKLIST

☐ ## IDENTIFY THE PROBLEM

In this section, dive deep into a possible **PROBLEM** you would like to solve. You might look at an object and ask yourself what more can be added to make it better? You may have a design **PROBLEM** that exists and you want to find a design solution for it. Or you may just want to invent something really cool! Once you determine a **PROBLEM** (**OPPORTUNITY** to create) you need to think about this further.

☐ ## NEEDS

After identifying your **PROBLEM (OPPORTUNITY)**, start to think about what the design you are creating **NEEDS** to have. These needs might be thought of as requirements. What does the design have to have in it? Such as: a certain size, characteristics, or something it must do.

☐ ## LIMITATIONS

Consider what **LIMITATIONS** you have in creating your design. Do you have a limited amount of time? Or only certain supplies or materials? This list will help you design within your limits while ensuring you include everything that is required.

☐ ## RESEARCH

RESEARCH to see what others have developed and created related to your **PROBLEM**. Here you might discover things you like about other people's designs and get inspiration for what you might include in your own design, along with things you might not include. You might also find things from other designs you do not want to include. Make notes on your research here.

☐ ## IMAGINE

During **IMAGINE** you will come up with ideas for your design. The goal here is to sketch out as many design ideas as possible, not worrying about whether the idea is good or bad.

☐ ## LIKES & DISLIKES

When you have imagined and sketched out as many ideas as you can, you will list the things you **LIKE** and **DISLIKE** about each your design ideas.

☐ ## FINAL DESIGN

Here you will come up with a **FINAL DESIGN**. Try to incorporate as many of the ideas you liked from the **IMAGINE** section. In your **FINAL DESIGN**, try to add in as much detail as possible, such as labels and explanations, so you know exactly what you want to create.

DESIGN PROCESS

IDENTIFY PROBLEM

RESEARCH

IMAGINE

DESIGN

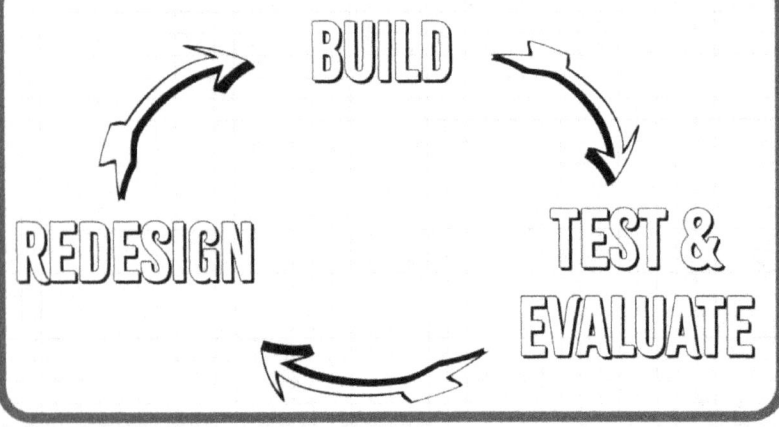

BUILD → TEST & EVALUATE → REDESIGN

SHOWCASE

> **Creative part of Design Thinking**
>
> In this section you will try to build the **FINAL DESIGN** you imagined. You will need to review any **LIMITATIONS** you have listed and consider the **NEEDS** (requirements) of the design. Even though you will be trying to build what you have imagined, there may be times when you might need to change this. Go gather your materials and start building **PROTOTYPE 1!**

 # 5 BUILD - CREATE YOUR DESIGN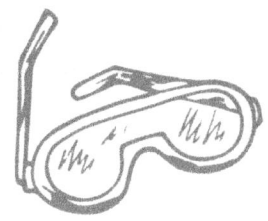

PROTOTYPE 1

**Try to build the FINAL DESIGN you created.
Take a picture of what you have created and add it here to remind you.**

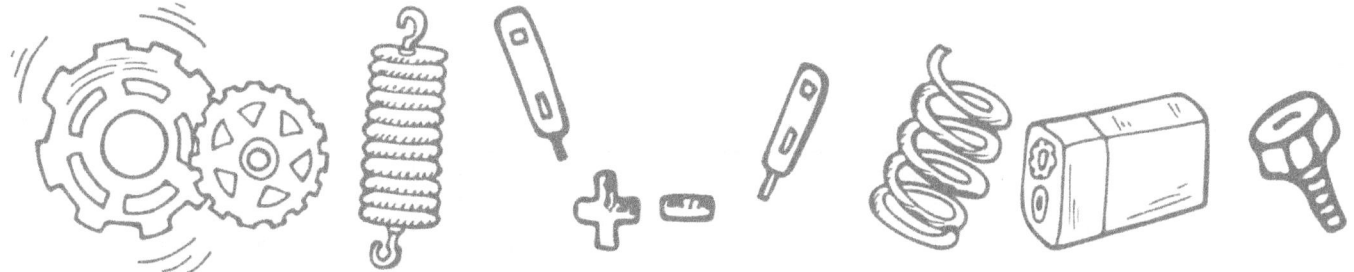

6 TEST & EVALUATE

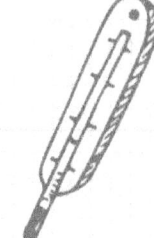

FEEDBACK & TEST

Show people your prototype and get **FEEDBACK** on what they like about it and what changes they might make to improve it. You may also have things you would like to change about your first prototype. You might need to **TEST** your prototype to see if it does what you planned. Write all of these ideas here so you can review them later when making improvements.

IMPROVE

Decide what changes you will make to your prototype. List what you want to **IMPROVE** with reasons as to why you want to make these changes. You might also decide to not change things, you can write these down as well.

 ☆

 ☆

 ☆

☆

5. BUILD (YES, STEP 5 AGAIN)

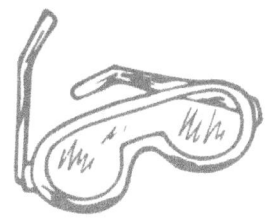

PROTOTYPE 2

Make changes to **PROTOTYPE 1** based on the feedback you received and the testing you conducted to create **PROTOTYPE 2**. Take a picture of **PROTOTYPE 2** when you are done and add it here.

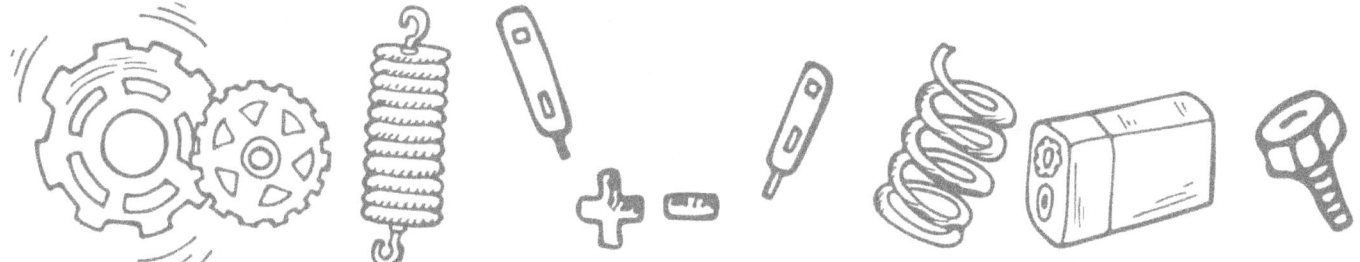

6 TEST & EVALUATE 2

FEEDBACK & TEST

Show people your prototype and get **FEEDBACK** on what they like about it and what changes they might make to improve it. You may also have things you would like to change about your prototype. You might need to **TEST** your protptype to see if it does what you planned. Write all of these ideas here so you can review them later when making improvements.

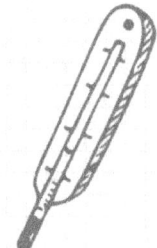

Extra Build, Test & Evaluate pages in the Appendix if you want to keep developing your prototype.

IMPROVE

Decide what changes you will make to your prototype. List what you want to **IMPROVE** with reasons as to why you want to make these changes. You might also decide to not change things, you can write these down as well.

CHECKLIST

☐ ### PROTOTYPE 1

Try to build the **FINAL DESIGN** you created.
Take a picture of what you created and add it here to remind you.

☐ ### FEEDBACK & TEST

Show people your prototype and get **FEEDBACK** on what they like about it and what changes they might make to it. You may have things you would like to change about what you have made. You might need to **TEST** your prototype to see if it does what you planned. Write all of these ideas here so you can review them later if you need to.

☐ ### IMPROVE

Decide what changes you will make to your prototype. Try to list what you want to **IMPROVE** with reasons as to why you want to make these changes. You might also decide to not change things, you can write these down as we

☐ ### PROTOTYPE 2

Make changes to **PROTOTYPE 1** based on the feedback you received and the testing you conducted to create **PROTOTYPE 2**. Take a picture of **PROTOTYPE 2** when you are done and add it here.

☐ ### THE DESIGN CYCLE

The **DESIGN CYCLE** can continue, by repeatedly gathering feedback, making changes to the prototype you create and then developing another prototype. This cycle can carry on until you are satisfied with the prototype you have created.

☐ ### APPENDIX

If you need to make more changes to your prototype as you **BUILD, TEST & EVALUATE** there are extra pages in the **APPENDIX** at the back of the book (pg.57).

DESIGN PROCESS

IDENTIFY PROBLEM

RESEARCH

IMAGINE

DESIGN

BUILD

REDESIGN  **TEST & EVALUATE**
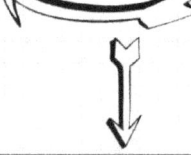
SHOWCASE

> ### Showcase part of Design Thinking
>
> In this final section you want to **SHOW+CASE** what you have created. In the **SHOWCASE** you will present your final design to a wider audience. Part of this section is the chance to **REFLECT** on the design thinking process you have completed. Thinking back on the process and the challenges faced, helps you learn about yourself for future design challenges.

 # SHOWCASE

TIME TO SHARE YOUR DESIGN

To SHOWCASE your project you can simply show a few people what you have created or you might want to present your design to a bigger audience. **A** class presentation, an exhibition or even an investment presentation, might be where you **SHOWCASE** your design!
Take a picture of what you have created and add it below.

 # REFLECT

What was your most enjoyable moment during the design thinking process?

Did anything unexpected happened? How did you respond to challenges?

What were your most interesting discoveries? This could be a new perspective, new technique or skill.

Describe any further changes you would make to your prototype if you had more time.

What did you learn about yourself during this process? OR How would you apply what you have learned in school, in your community or with friends and your family?

DESIGN NO. 2

DESIGN PROCESS

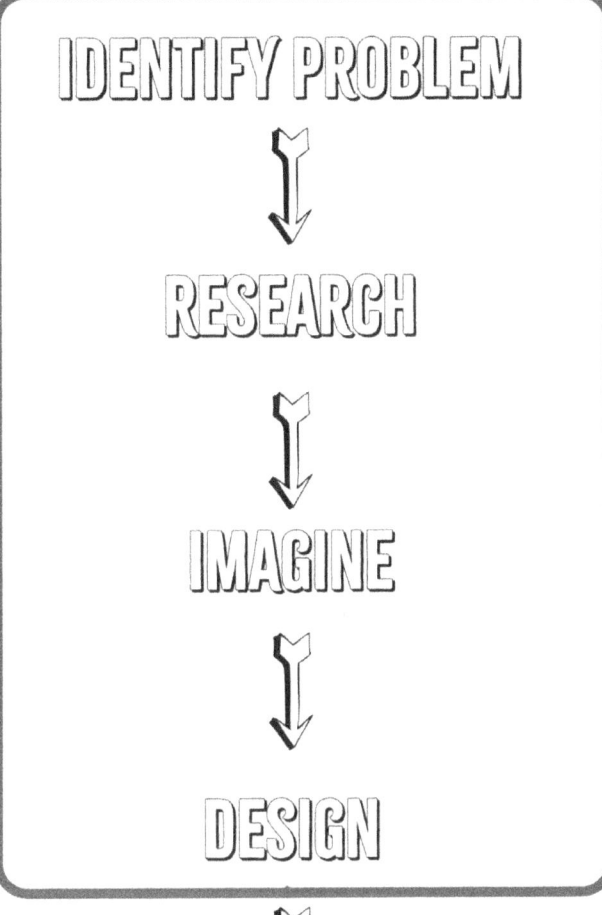

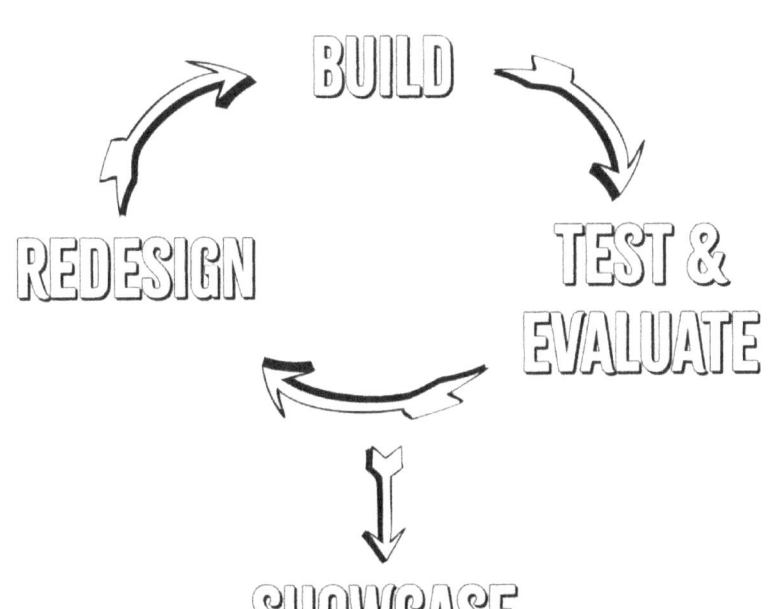

Exploration part of Design Thinking

In this section, dive deep into a possible **PROBLEM** you would like to solve. You might look at an object and ask yourself what more can be added to make it better? You may have a design **PROBLEM** that exists and you want to find a design solution for it. Or you may just want to invent something really cool! Once you determine a **PROBLEM** (**OPPORTUNITY** to create) you need to think about this further.

 # 1 IDENTIFY THE PROBLEM

IDENTIFY THE PROBLEM
Identify the problem you want to solve.

NEEDS
After identifying your **PROBLEM (OPPORTUNITY)**, start to think about what the design (solution) you are creating **NEEDS** to have. These needs might be thought of as requirements. What does the design have to have in it? Such as: a certain size, characteristics, or something it must do.

LIMITATIONS
Consider what **LIMITATIONS** you have in creating your design. Do you have a limited amount of time? Or only certain supplies or materials? This list will help you design within your limits while ensuring you include everything that is required.

RESEARCH

WHAT DID YOU FIND OUT?

RESEARCH to see what others have developed and created related to your **PROBLEM**. Here you might discover things you like about other people's designs and get inspiration for what you might include in your own design. You might also find things from other designs you do not want to include. Make notes on your research here.

3 IMAGINE, IMAGINE..

During **IMAGINE** you will come up with ideas for your design. The goal here is to sketch out as many design ideas as possible, not worrying about whether the idea is good or bad.

When you have imagined and sketched out as many ideas as you can, you will list the things you **LIKE** and **DISLIKE** about each your design ideas. Try comparing your ideas to come up with these.

IDEA 1:

IDEA 2:

LIKES:

DISLIKES:

LIKES:

DISLIKES:

..... IMAGINE!

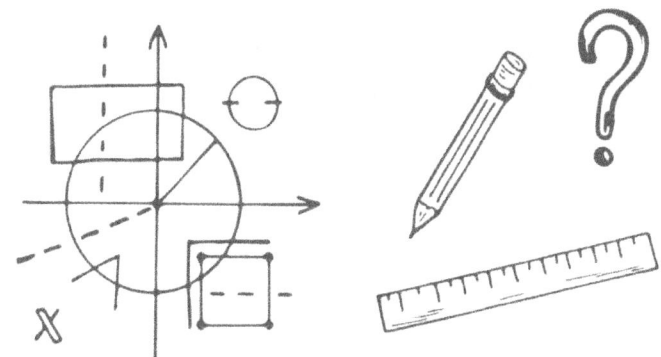

IDEA 3:

IDEA 4:

LIKES:

DISLIKES:

LIKES:

DISLIKES:

4 FINAL DESIGN

FINAL DESIGN

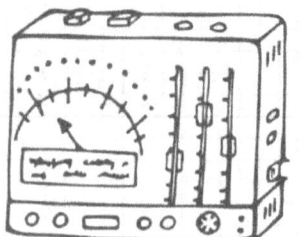

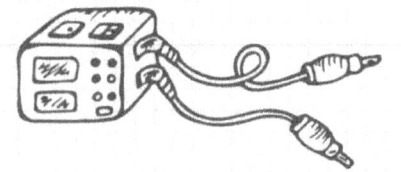

Here you will come up with a **FINAL DESIGN**. Try to incorporate as many of the ideas you liked from the **IMAGINE** section. In your **FINAL DESIGN**, try to add in as much detail as possible, such as labels and explanations, so you know exactly what you want to create and build.

CHECKLIST

- [ ] **IDENTIFY THE PROBLEM**

    In this section, dive deep into a possible **PROBLEM** you would like to solve. You might look at an object and ask yourself what more can be added to make it better? You may have a design **PROBLEM** that exists and you want to find a design solution for it. Or you may just want to invent something really cool! Once you determine a **PROBLEM (OPPORTUNITY** to create) you need to think about this further.

- [ ] **NEEDS**

    After identifying your **PROBLEM (OPPORTUNITY)**, start to think about what the design you are creating **NEEDS** to have. These needs might be thought of as requirements. What does the design have to have in it? Such as: a certain size, characteristics, or something it must do.

- [ ] **LIMITATIONS**

    Consider what **LIMITATIONS** you have in creating your design. Do you have a limited amount of time? Or only certain supplies or materials? This list will help you design within your limits while ensuring you include everything that is required.

- [ ] **RESEARCH**

    RESEARCH to see what others have developed and created related to your **PROBLEM**. Here you might discover things you like about other people's designs and get inspiration for what you might include in your own design, along with things you might not include. You might also find things from other designs you do not want to include. Make notes on your research here.

- [ ] **IMAGINE**

    During **IMAGINE** you will come up with ideas for your design. The goal here is to sketch out as many design ideas as possible, not worrying about whether the idea is good or bad.

- [ ] **LIKES & DISLIKES**

    When you have imagined and sketched out as many ideas as you can, you will list the things you **LIKE** and **DISLIKE** about each your design ideas.

- [ ] **FINAL DESIGN**

    Here you will come up with a **FINAL DESIGN**. Try to incorporate as many of the ideas you liked from the **IMAGINE** section. In your **FINAL DESIGN**, try to add in as much detail as possible, such as labels and explanations, so you know exactly what you want to create.

DESIGN PROCESS

IDENTIFY PROBLEM

RESEARCH

IMAGINE

DESIGN

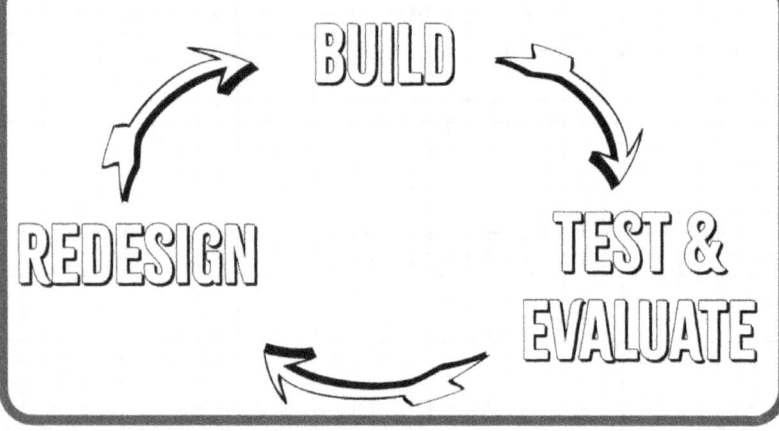

SHOWCASE

> **Creative part of Design Thinking**
>
> In this section you will try to build the **FINAL DESIGN** you imagined. You will need to review any **LIMITATIONS** you have listed and consider the **NEEDS** (requirements) of the design. Even though you will be trying to build what you have imagined, there may be times when you might need to change this. Go gather your materials and start building **PROTOTYPE 1**!

5. BUILD - CREATE YOUR DESIGN

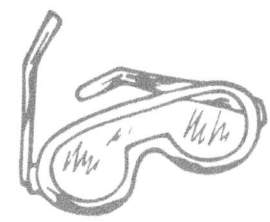

PROTOTYPE 1

Try to build the FINAL DESIGN you created.
Take a picture of what you have created and add it here to remind you.

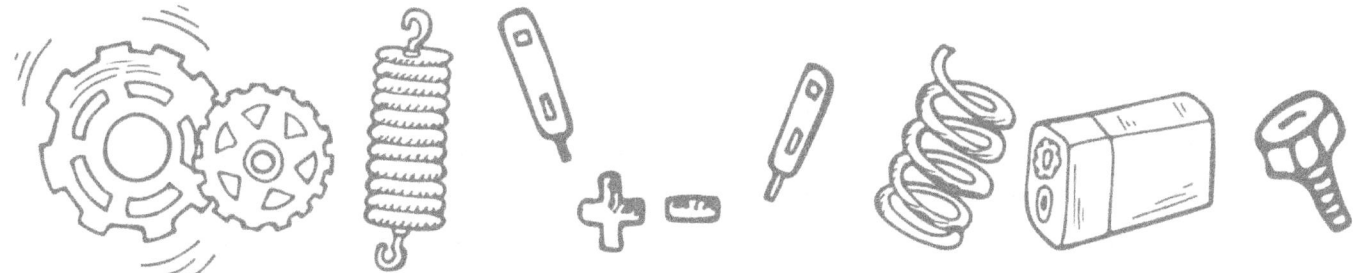

 # TEST & EVALUATE
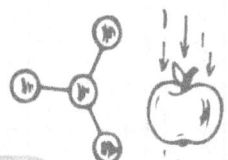

FEEDBACK & TEST

Show people your prototype and get **FEEDBACK** on what they like about it and what changes they might make to improve it. You may also have things you would like to change about your first prototype. You might need to **TEST** your prototype to see if it does what you planned. Write all of these ideas here so you can review them later when making improvements.

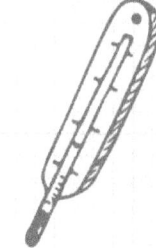

IMPROVE

Decide what changes you will make to your prototype. List what you want to **IMPROVE** with reasons as to why you want to make these changes. You might also decide to not change things, you can write these down as well.

☆

☆

☆

☆

5. BUILD (YES, STEP 5 AGAIN)

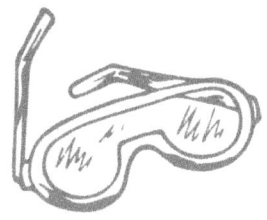

PROTOTYPE 2

Make changes to **PROTOTYPE 1** based on the feedback you received and the testing you conducted to create **PROTOTYPE 2**. Take a picture of **PROTOTYPE 2** when you are done and add it here.

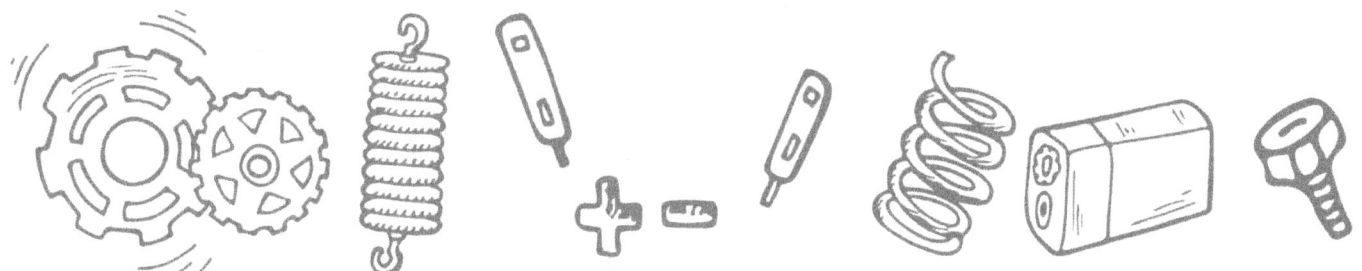

 # 6 TEST & EVALUATE 2

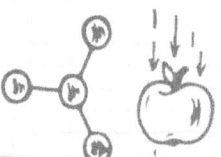

FEEDBACK & TEST

Show people your prototype and get **FEEDBACK** on what they like about it and what changes they might make to improve it. You may also have things you would like to change about your prototype. You might need to **TEST** your protptype to see if it does what you planned. Write all of these ideas here so you can review them later when making improvements.

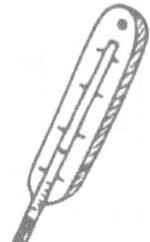

Extra Build, Test & Evaluate pages in the Appendix if you want to keep developing your prototype.

IMPROVE
Decide what changes you will make to your prototype. List what you want to **IMPROVE** with reasons as to why you want to make these changes. You might also decide to not change things, you can write these down as well.

☆

☆

☆

☆

CHECKLIST

☐ ### PROTOTYPE 1

Try to build the **FINAL DESIGN** you created.
Take a picture of what you created and add it here to remind you.

☐ ### FEEDBACK & TEST

Show people your prototype and get **FEEDBACK** on what they like about it and what changes they might make to it. You may have things you would like to change about what you have made. You might need to **TEST** your prototype to see if it does what you planned. Write all of these ideas here so you can review them later if you need to.

☐ ### IMPROVE

Decide what changes you will make to your prototype. Try to list what you want to **IMPROVE** with reasons as to why you want to make these changes. You might also decide to not change things, you can write these down as we

☐ ### PROTOTYPE 2

Make changes to **PROTOTYPE 1** based on the feedback you received and the testing you conducted to create **PROTOTYPE 2**. Take a picture of **PROTOTYPE 2** when you are done and add it here.

☐ ### THE DESIGN CYCLE

The **DESIGN CYCLE** can continue, by repeatedly gathering feedback, making changes to the prototype you create and then developing another prototype. This cycle can carry on until you are satisfied with the prototype you have created.

☐ ### APPENDIX

If you need to make more changes to your prototype as you **BUILD, TEST & EVALUATE** there are extra pages in the **APPENDIX** at the back of the book (pg.57).

DESIGN PROCESS

IDENTIFY PROBLEM

RESEARCH

IMAGINE

DESIGN

BUILD

REDESIGN  **TEST & EVALUATE**

SHOWCASE

> **Showcase part of Design Thinking**
>
> In this final section you want to **SHOW+CASE** what you have created. In the **SHOWCASE** you will present your final design to a wider audience. Part of this section is the chance to **REFLECT** on the design thinking process you have completed. Thinking back on the process and the challenges faced, helps you learn about yourself for future design challenges.

TIME TO SHARE YOUR DESIGN

To SHOWCASE your project you can simply show a few people what you have created or you might want to present your design to a bigger audience. A class presentation, an exhibition or even an investment presentation, might be where you **SHOWCASE** your design!
Take a picture of what you have created and add it below.

 # REFLECT

What was your most enjoyable moment during the design thinking process?

Did anything unexpected happened? How did you respond to challenges?

What were your most interesting discoveries? This could be a new perspective, new technique or skill.

Describe any further changes you would make to your prototype if you had more time.

What did you learn about yourself during this process? OR How would you apply what you have learned in school, in your community or with friends and your family?

DESIGN NO.3

DESIGN PROCESS

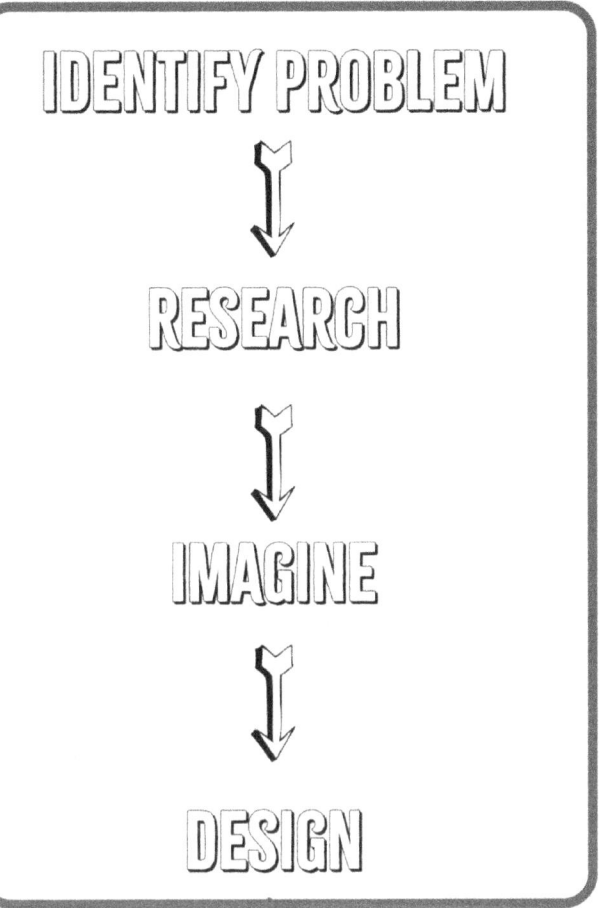

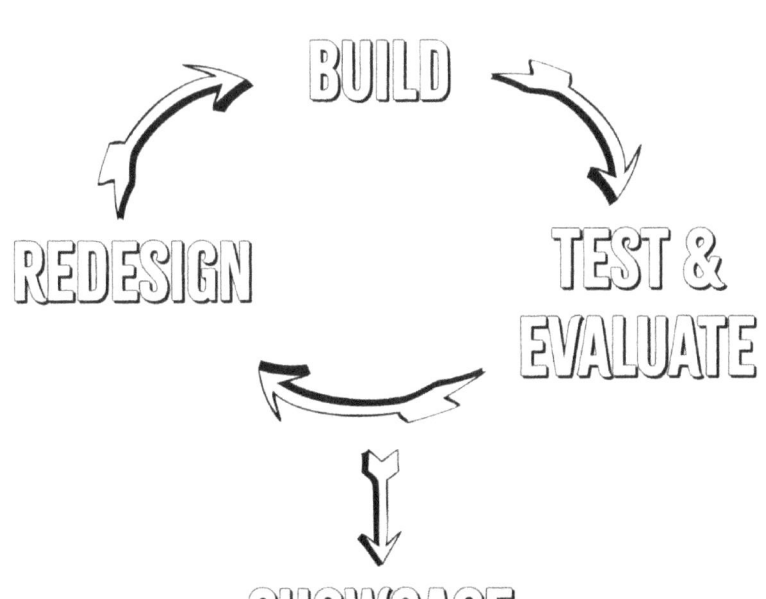

Exploration part of Design Thinking

In this section, dive deep into a possible **PROBLEM** you would like to solve. You might look at an object and ask yourself what more can be added to make it better? You may have a design **PROBLEM** that exists and you want to find a design solution for it. Or you may just want to invent something really cool! Once you determine a **PROBLEM** (**OPPORTUNITY** to create) you need to think about this further.

 # 1 IDENTIFY THE PROBLEM

IDENTIFY THE PROBLEM

Identify the problem you want to solve.

 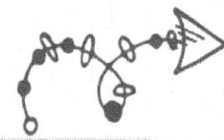

NEEDS

After identifying your **PROBLEM (OPPORTUNITY)**, start to think about what the design (solution) you are creating **NEEDS** to have. These needs might be thought of as requirements. What does the design have to have in it? Such as: a certain size, characteristics, or something it must do.

LIMITATIONS

Consider what **LIMITATIONS** you have in creating your design. Do you have a limited amount of time? Or only certain supplies or materials? This list will help you design within your limits while ensuring you include everything that is required.

RESEARCH

WHAT DID YOU FIND OUT?

RESEARCH to see what others have developed and created related to your **PROBLEM**. Here you might discover things you like about other people's designs and get inspiration for what you might include in your own design. You might also find things from other designs you do not want to include. Make notes on your research here.

3. IMAGINE, IMAGINE....

During **IMAGINE** you will come up with ideas for your design. The goal here is to sketch out as many design ideas as possible, not worrying about whether the idea is good or bad.

When you have imagined and sketched out as many ideas as you can, you will list the things you **LIKE** and **DISLIKE** about each your design ideas. Try comparing your ideas to come up with these.

IDEA 1:

IDEA 2:

LIKES:

DISLIKES:

LIKES:

DISLIKES:

..... IMAGINE!

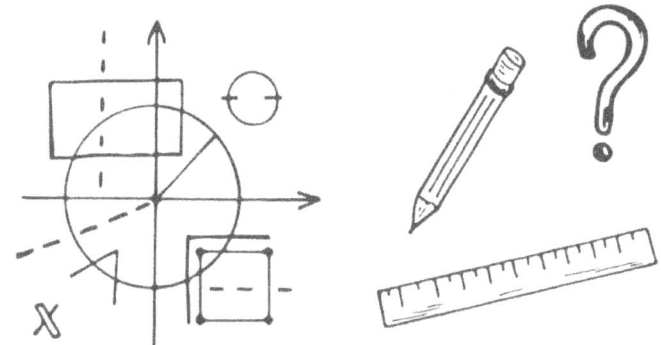

IDEA 3:

IDEA 4:

LIKES:

DISLIKES:

LIKES:

DISLIKES:

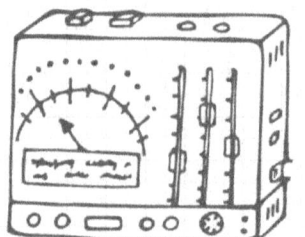

FINAL DESIGN

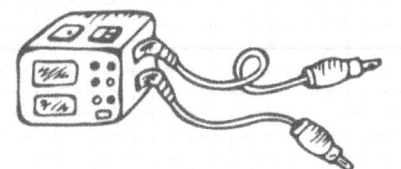

Here you will come up with a **FINAL DESIGN**. Try to incorporate as many of the ideas you liked from the **IMAGINE** section. In your **FINAL DESIGN**, try to add in as much detail as possible, such as labels and explanations, so you know exactly what you want to create and build.

CHECKLIST

- ### IDENTIFY THE PROBLEM
  In this section, dive deep into a possible **PROBLEM** you would like to solve. You might look at an object and ask yourself what more can be added to make it better? You may have a design **PROBLEM** that exists and you want to find a design solution for it. Or you may just want to invent something really cool! Once you determine a **PROBLEM** (**OPPORTUNITY** to create) you need to think about this further.

- ### NEEDS
  After identifying your **PROBLEM** (**OPPORTUNITY**), start to think about what the design you are creating **NEEDS** to have. These needs might be thought of as requirements. What does the design have to have in it? Such as: a certain size, characteristics, or something it must do.

- ### LIMITATIONS
  Consider what **LIMITATIONS** you have in creating your design. Do you have a limited amount of time? Or only certain supplies or materials? This list will help you design within your limits while ensuring you include everything that is required.

- ### RESEARCH
  RESEARCH to see what others have developed and created related to your **PROBLEM**. Here you might discover things you like about other people's designs and get inspiration for what you might include in your own design, along with things you might not include. You might also find things from other designs you do not want to include. Make notes on your research here.

- ### IMAGINE
  During **IMAGINE** you will come up with ideas for your design. The goal here is to sketch out as many design ideas as possible, not worrying about whether the idea is good or bad.

- ### LIKES & DISLIKES
  When you have imagined and sketched out as many ideas as you can, you will list the things you **LIKE** and **DISLIKE** about each your design ideas.

- ### FINAL DESIGN
  Here you will come up with a **FINAL DESIGN**. Try to incorporate as many of the ideas you liked from the **IMAGINE** section. In your **FINAL DESIGN**, try to add in as much detail as possible, such as labels and explanations, so you know exactly what you want to create.

DESIGN PROCESS

IDENTIFY PROBLEM

RESEARCH

IMAGINE

DESIGN

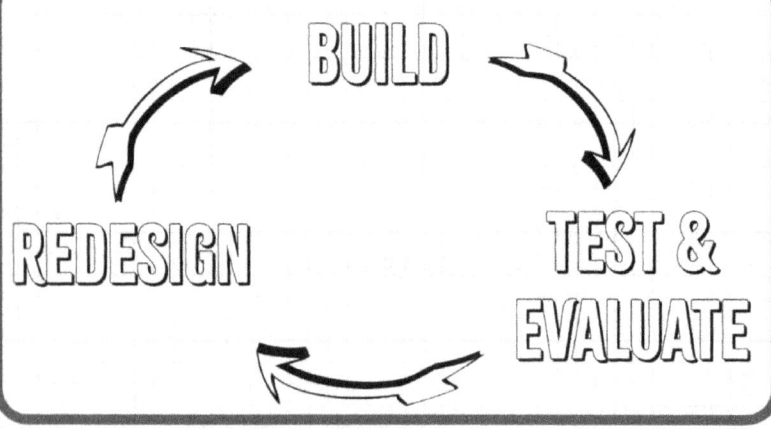

BUILD → TEST & EVALUATE → REDESIGN → (BUILD)

SHOWCASE

Creative part of Design Thinking

In this section you will try to build the **FINAL DESIGN** you imagined. You will need to review any **LIMITATIONS** you have listed and consider the **NEEDS** (requirements) of the design. Even though you will be trying to build what you have imagined, there may be times when you might need to change this. Go gather your materials and start building **PROTOTYPE 1**!

5 BUILD - CREATE YOUR DESIGN

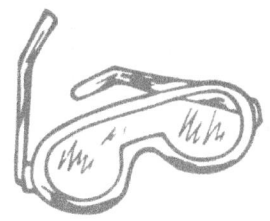

PROTOTYPE 1

Try to build the FINAL DESIGN you created.
Take a picture of what you have created and add it here to remind you.

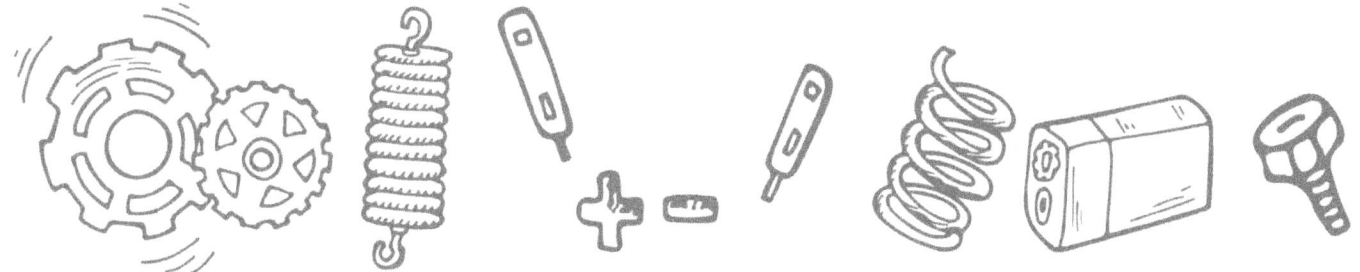

# 6 TEST & EVALUATE

FEEDBACK & TEST

Show people your prototype and get **FEEDBACK** on what they like about it and what changes they might make to improve it. You may also have things you would like to change about your first prototype. You might need to **TEST** your prototype to see if it does what you planned. Write all of these ideas here so you can review them later when making improvements.

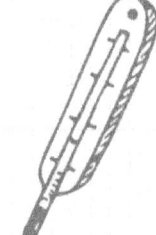

IMPROVE

Decide what changes you will make to your prototype. List what you want to **IMPROVE** with reasons as to why you want to make these changes. You might also decide to not change things, you can write these down as well.

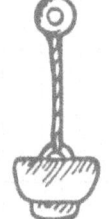

☆

☆

☆

☆

5. BUILD (YES, STEP 5 AGAIN)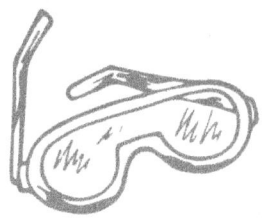

PROTOTYPE 2

Make changes to **PROTOTYPE 1** based on the feedback you received and the testing you conducted to create **PROTOTYPE 2**. Take a picture of **PROTOTYPE 2** when you are done and add it here.

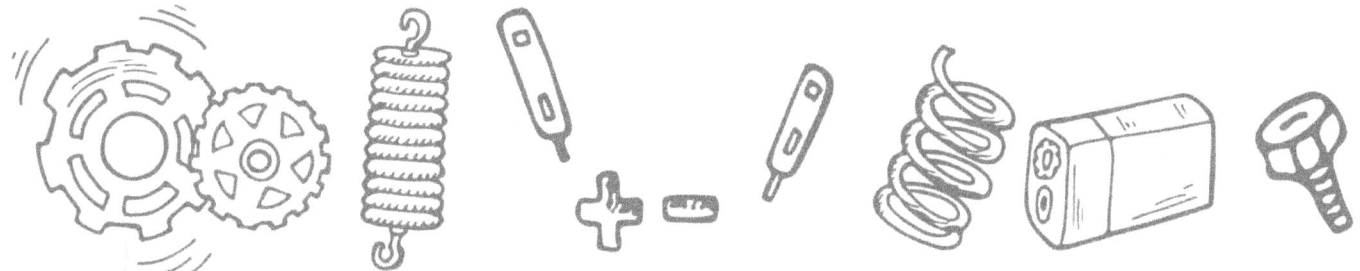

# 6 TEST & EVALUATE 2

FEEDBACK & TEST

Show people your prototype and get **FEEDBACK** on what they like about it and what changes they might make to improve it. You may also have things you would like to change about your prototype. You might need to **TEST** your protptype to see if it does what you planned. Write all of these ideas here so you can review them later when making improvements.

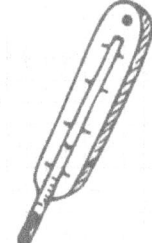

Extra Build, Test & Evaluate pages in the Appendix if you want to keep developing your prototype.

IMPROVE

Decide what changes you will make to your prototype. List what you want to **IMPROVE** with reasons as to why you want to make these changes. You might also decide to not change things, you can write these down as well.

☆

☆

☆

☆

CHECKLIST

- [ ] ### PROTOTYPE 1
  Try to build the **FINAL DESIGN** you created.
  Take a picture of what you created and add it here to remind you.

- [ ] ### FEEDBACK & TEST
  Show people your prototype and get **FEEDBACK** on what they like about it and what changes they might make to it. You may have things you would like to change about what you have made. You might need to **TEST** your prototype to see if it does what you planned. Write all of these ideas here so you can review them later if you need to.

- [ ] ### IMPROVE
  Decide what changes you will make to your prototype. Try to list what you want to **IMPROVE** with reasons as to why you want to make these changes. You might also decide to not change things, you can write these down as we

- [ ] ### PROTOTYPE 2
  Make changes to **PROTOTYPE 1** based on the feedback you received and the testing you conducted to create **PROTOTYPE 2**. Take a picture of **PROTOTYPE 2** when you are done and add it here.

- [ ] ### THE DESIGN CYCLE
  The **DESIGN CYCLE** can continue, by repeatedly gathering feedback, making changes to the prototype you create and then developing another prototype. This cycle can carry on until you are satisfied with the prototype you have created.

- [ ] ### APPENDIX
  If you need to make more changes to your prototype as you **BUILD, TEST & EVALUATE** there are extra pages in the **APPENDIX** at the back of the book (pg.57).

DESIGN PROCESS

IDENTIFY PROBLEM

RESEARCH

IMAGINE

DESIGN

BUILD

TEST & EVALUATE
REDESIGN
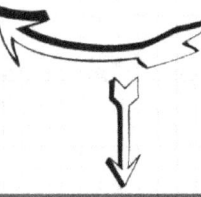
SHOWCASE

Showcase part of Design Thinking

In this final section you want to **SHOW+CASE** what you have created. In the **SHOWCASE** you will present your final design to a wider audience. Part of this section is the chance to **REFLECT** on the design thinking process you have completed. Thinking back on the process and the challenges faced, helps you learn about yourself for future design challenges.

7 SHOWCASE

TIME TO SHARE YOUR DESIGN

To SHOWCASE your project you can simply show a few people what you have created or you might want to present your design to a bigger audience. A class presentation, an exhibition or even an investment presentation, might be where you **SHOWCASE** your design!
Take a picture of what you have created and add it below.

 # REFLECT

What was your most enjoyable moment during the design thinking process?

Did anything unexpected happened? How did you respond to challenges?

What were your most interesting discoveries? This could be a new perspective, new technique or skill.

Describe any further changes you would make to your prototype if you had more time.

What did you learn about yourself during this process? OR How would you apply what you have learned in school, in your community or with friends and your family?

APPENDIX

Extra Build, Test & Evaluate Pages

 # BUILD - CREATE YOUR DESIGN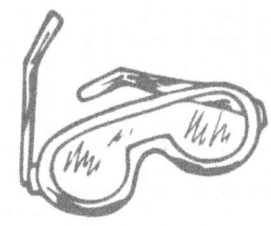

PROTOTYPE

Take a picture of what you created and add it here to remind you.

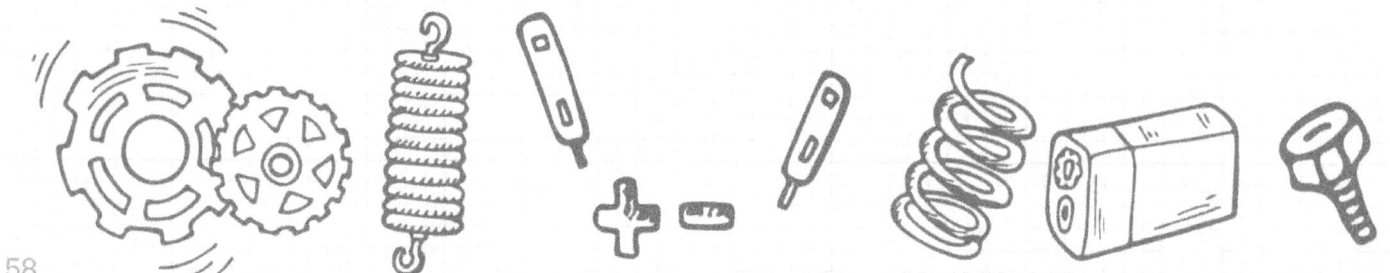

 # TEST & EVALUATE

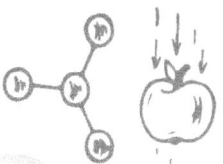

FEEDBACK & TEST

Show people your prototype and get **FEEDBACK** on what they like about it and what changes they might make to improve it. You may also have things you would like to change about your first prototype. You might need to **TEST** your prototype to see if it does what you planned. Write all of these ideas here so you can review them later when making improvements.

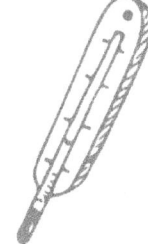

IMPROVE

Decide what changes you will make to your prototype. List what you want to **IMPROVE** with reasons as to why you want to make these changes. You might also decide to not change things, you can write these down as well.

5 BUILD - CREATE YOUR DESIGN
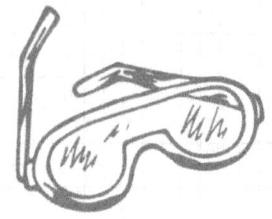

PROTOTYPE

Take a picture of what you created and add it here to remind you.

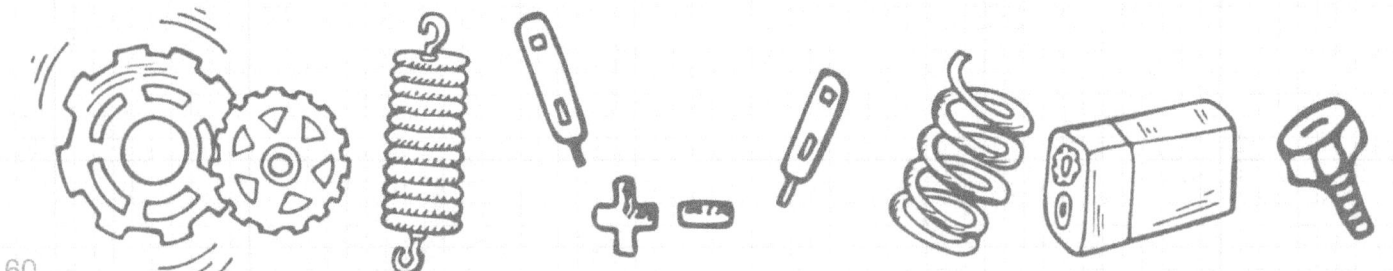

6 TEST & EVALUATE

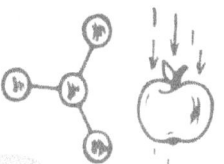

FEEDBACK & TEST

Show people your prototype and get **FEEDBACK** on what they like about it and what changes they might make to improve it. You may also have things you would like to change about your first prototype. You might need to **TEST** your prototype to see if it does what you planned. Write all of these ideas here so you can review them later when making improvements.

IMPROVE

Decide what changes you will make to your prototype. List what you want to **IMPROVE** with reasons as to why you want to make these changes. You might also decide to not change things, you can write these down as well.

☆

☆

☆

☆

 # 5 BUILD - CREATE YOUR DESIGN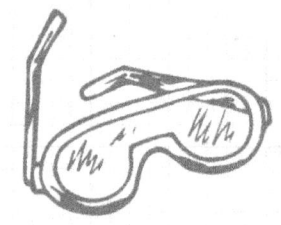

PROTOTYPE

Take a picture of what you created and add it here to remind you.

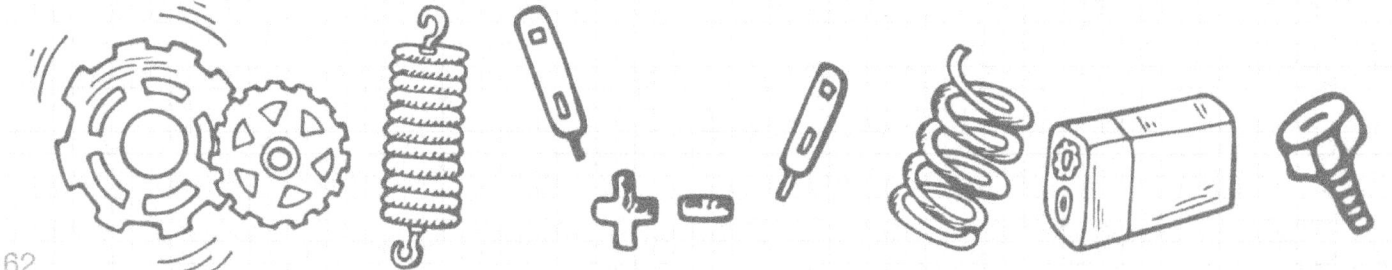

 # TEST & EVALUATE

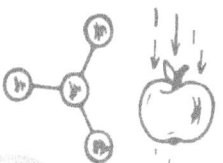

FEEDBACK & TEST

Show people your prototype and get **FEEDBACK** on what they like about it and what changes they might make to improve it. You may also have things you would like to change about your first prototype. You might need to **TEST** your prototype to see if it does what you planned. Write all of these ideas here so you can review them later when making improvements.

 **IMPROVE** Decide what changes you will make to your prototype. List what you want to **IMPROVE** with reasons as to why you want to make these changes. You might also decide to not change things, you can write these down as well.

☆

☆

☆

☆

BUILD - CREATE YOUR DESIGN

PROTOTYPE

Take a picture of what you created and add it here to remind you.

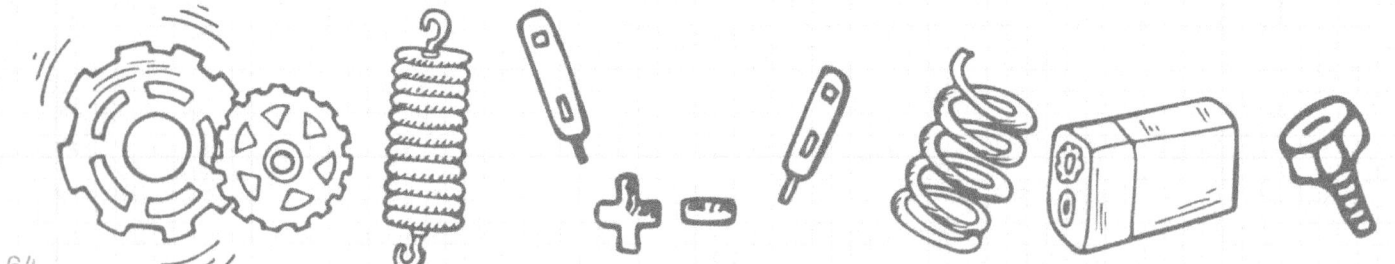

# 6 TEST & EVALUATE

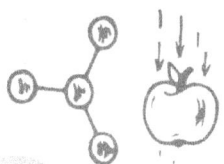

FEEDBACK & TEST

Show people your prototype and get **FEEDBACK** on what they like about it and what changes they might make to improve it. You may also have things you would like to change about your first prototype. You might need to **TEST** your prototype to see if it does what you planned. Write all of these ideas here so you can review them later when making improvements.

IMPROVE

Decide what changes you will make to your prototype. List what you want to **IMPROVE** with reasons as to why you want to make these changes. You might also decide to not change things, you can write these down as well.

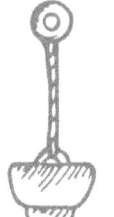

☆

☆

☆

☆

5 BUILD - CREATE YOUR DESIGN

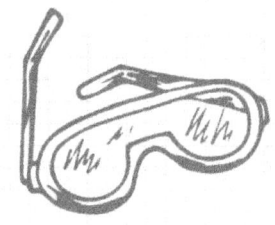

PROTOTYPE

Take a picture of what you created and add it here to remind you.

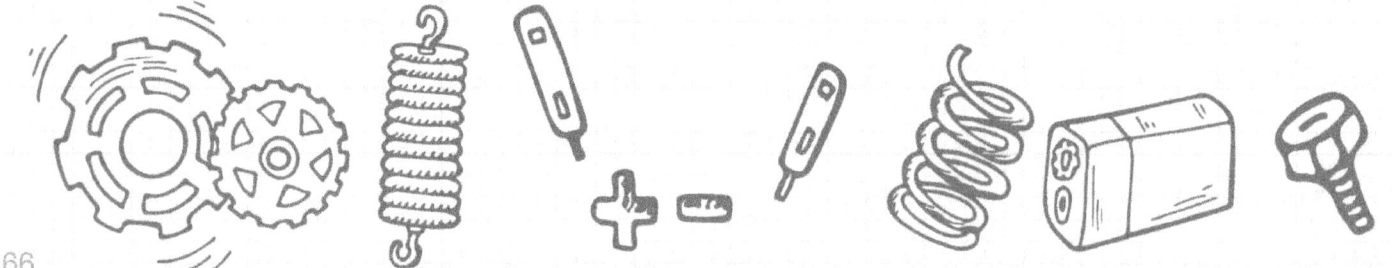

 # 6 TEST & EVALUATE

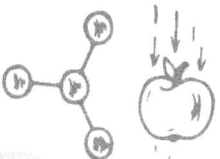
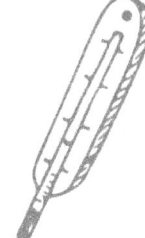

FEEDBACK & TEST

Show people your prototype and get **FEEDBACK** on what they like about it and what changes they might make to improve it. You may also have things you would like to change about your first prototype. You might need to **TEST** your prototype to see if it does what you planned. Write all of these ideas here so you can review them later when making improvements.

IMPROVE

Decide what changes you will make to your prototype. List what you want to **IMPROVE** with reasons as to why you want to make these changes. You might also decide to not change things, you can write these down as well.

☆

5 BUILD - CREATE YOUR DESIGN

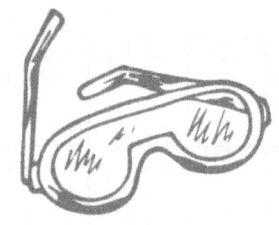

PROTOTYPE

Take a picture of what you created and add it here to remind you.

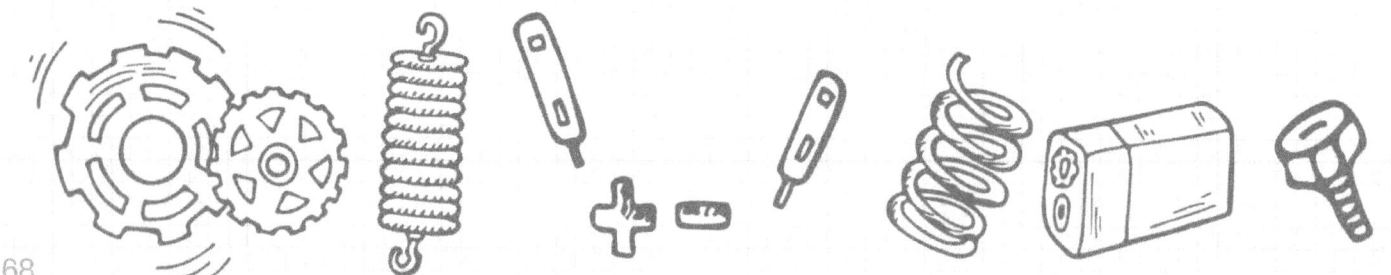

# TEST & EVALUATE

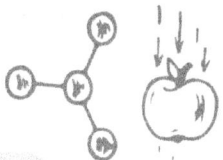

FEEDBACK & TEST

Show people your prototype and get **FEEDBACK** on what they like about it and what changes they might make to improve it. You may also have things you would like to change about your first prototype. You might need to **TEST** your prototype to see if it does what you planned. Write all of these ideas here so you can review them later when making improvements.

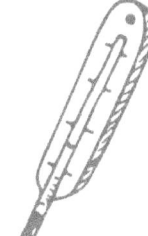

IMPROVE

Decide what changes you will make to your prototype. List what you want to **IMPROVE** with reasons as to why you want to make these changes. You might also decide to not change things, you can write these down as well.

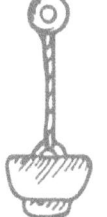

5 BUILD - CREATE YOUR DESIGN
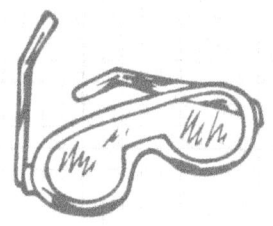

PROTOTYPE

Take a picture of what you created and add it here to remind you.

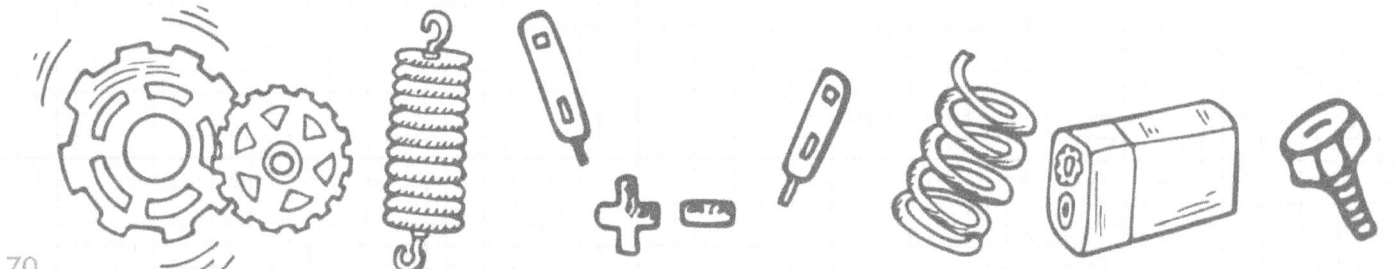

 # TEST & EVALUATE

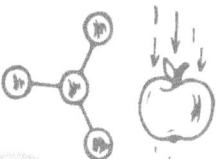

FEEDBACK & TEST

Show people your prototype and get **FEEDBACK** on what they like about it and what changes they might make to improve it. You may also have things you would like to change about your first prototype. You might need to **TEST** your prototype to see if it does what you planned. Write all of these ideas here so you can review them later when making improvements.

IMPROVE

Decide what changes you will make to your prototype. List what you want to **IMPROVE** with reasons as to why you want to make these changes. You might also decide to not change things, you can write these down as well.

DESIGN PROCESS

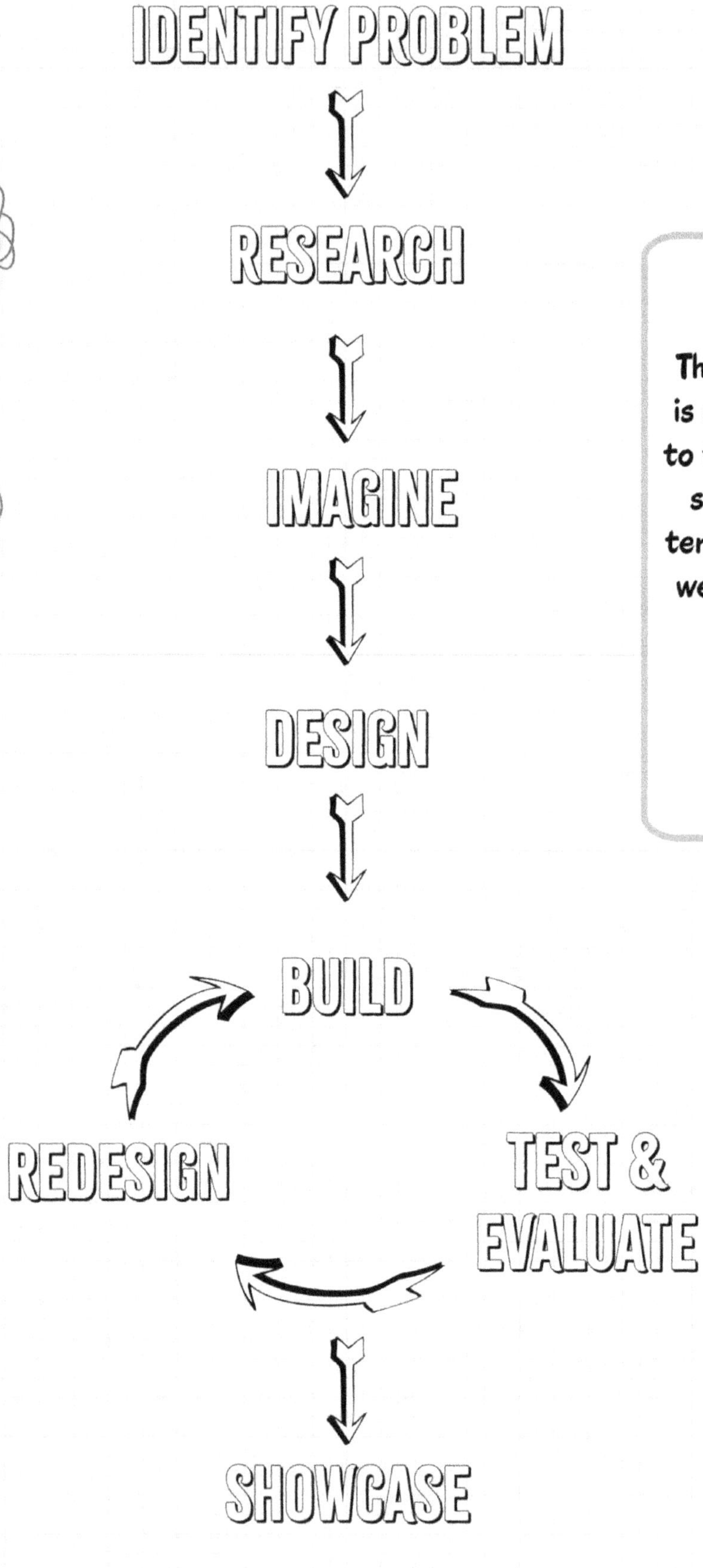

IDENTIFY PROBLEM
⬇
RESEARCH
⬇
IMAGINE
⬇
DESIGN
⬇
BUILD ⟶ **TEST & EVALUATE** ⟶ **REDESIGN** ⟶ (back to BUILD)
⬇
SHOWCASE

> The design thinking process is an easy to use framework to facilitate creative problem solving. By using planning templates to generate ideas, we hope to help you to find solutions to real-world design challenges.

Notes